Understanding M

A Level Pl

David Drumm

Before we start

If you do A level maths as well as A level physics you will have done a fair bit of mechanics and you may feel that you have this topic sussed. For example most maths specifications include the use of suvat equations. You should be careful however for several reasons:

- You will not cover all the topics needed for the A level physics requirement
- In physics the theory is very much based in real life contexts. In maths it can be very analytical without much discussion of practical uses.
- The level at which the material is presented is often different and often the mathematical requirements differ.

My goal is to start with the basic concepts then work up to more advanced situations. Much of the material in here builds on GCSE physics work so unless that is well understood then misconceptions can persist. So many year 12 students come to me not understanding the difference between mass and weight!

Do not look at a section without fully studying the sections before it.

I hope you enjoy studying this topic and it improves your understanding of many fundamental concepts which will crop up again and again throughout your physics course.

Displacement time graphs

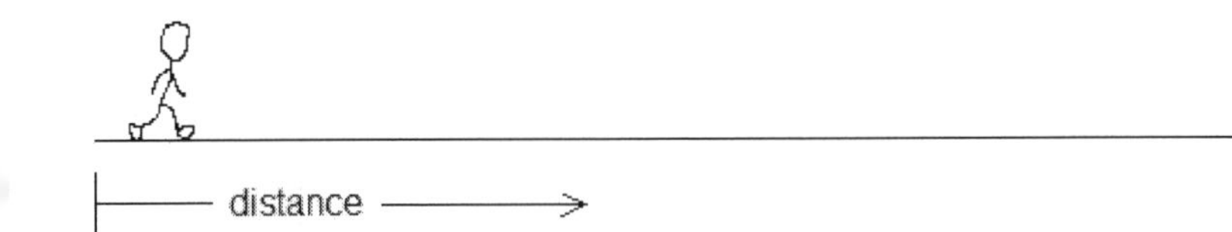

Imagine you walked in a straight line and every 2 seconds you dropped a marker on the floor. You could then measure how far you had walked at different times and plot a displacement time graph.

Note that displacement, as a vector quantity, is the distance travelled **in a certain direction**. We will discuss vectors at length later.

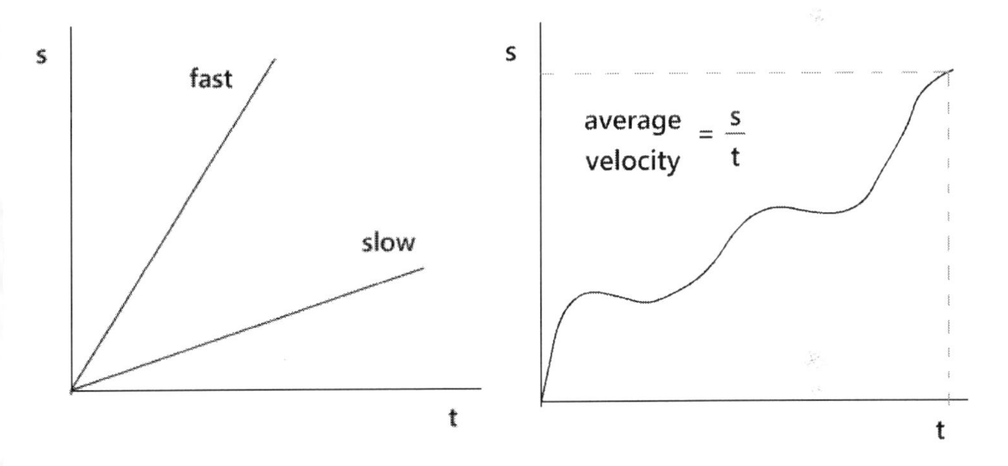

Above is a displacement time graph for a slow walk and a fast walk. We shall use the symbol “s” for displacement.

When you travel faster the graph is steeper as your velocity, your speed in a certain direction, is greater.

If your velocity changes during the journey then calculating $\frac{s}{t}$ will tell you your average velocity for the journey.

This is different to your **instantaneous velocity** at any point in time. To find that you would find the gradient of a tangent at the corresponding point on the graph. Instantaneous velocity is expressed as $v = \frac{ds}{dt}$. It is the rate of change of displacement at an instant in time.

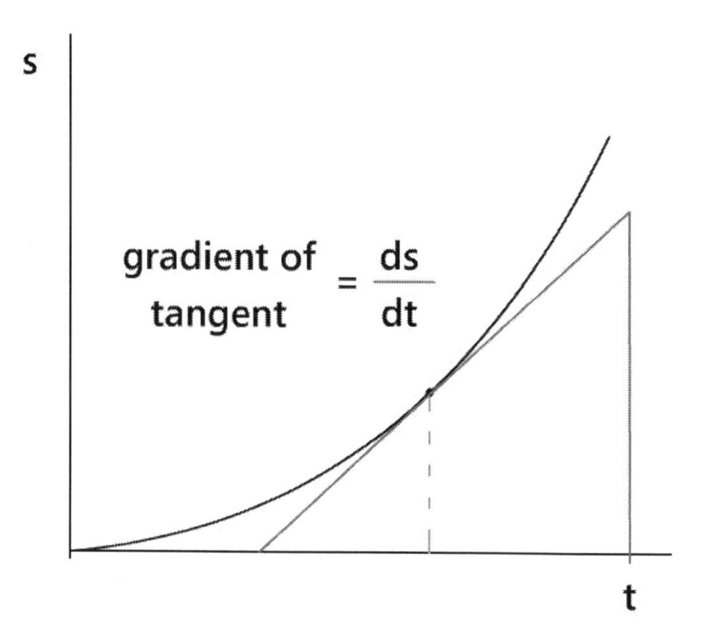

average velocity = $\frac{s}{t}$ instantaneous velocity = $\frac{ds}{dt}$

Velocity time graphs

Imagine you went for a journey in a car. If the journey is down a straight road then your velocity will be your speed in that direction. Imagine you made a note of your velocity every five seconds then plotted a graph. Below are a few possible graphs.

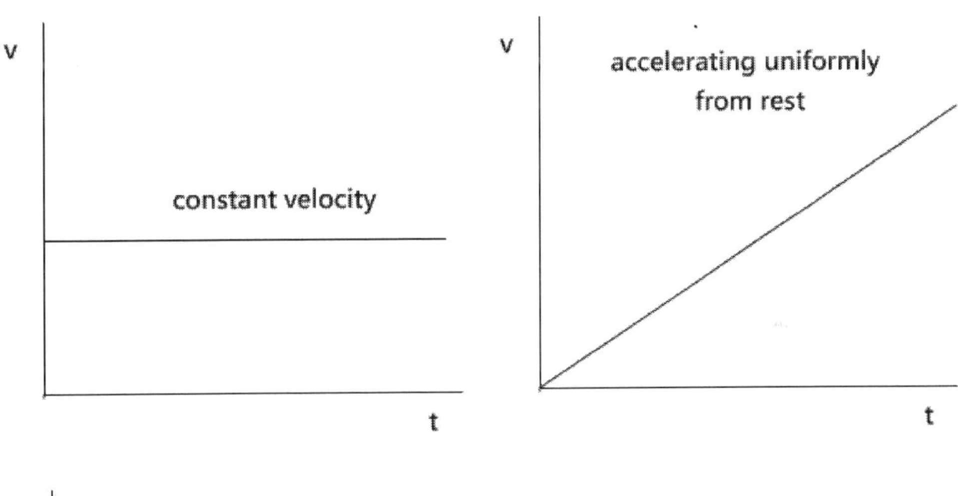

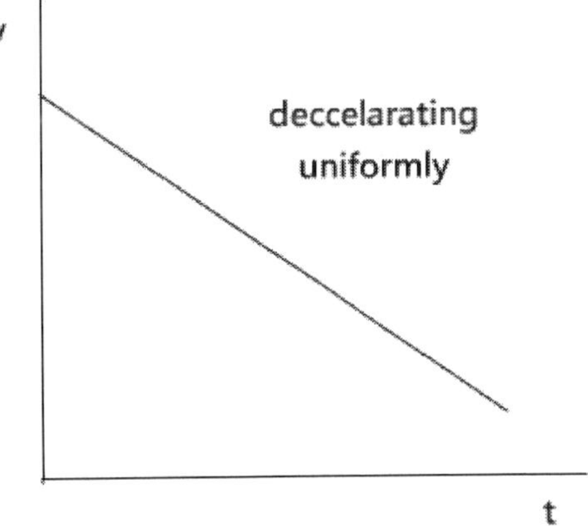

Displacement from a velocity time graph

If you were travelling at a constant velocity of 12m/s for 20s how far would you travel?

The answer will be 240m as 12 x 20 = 240.

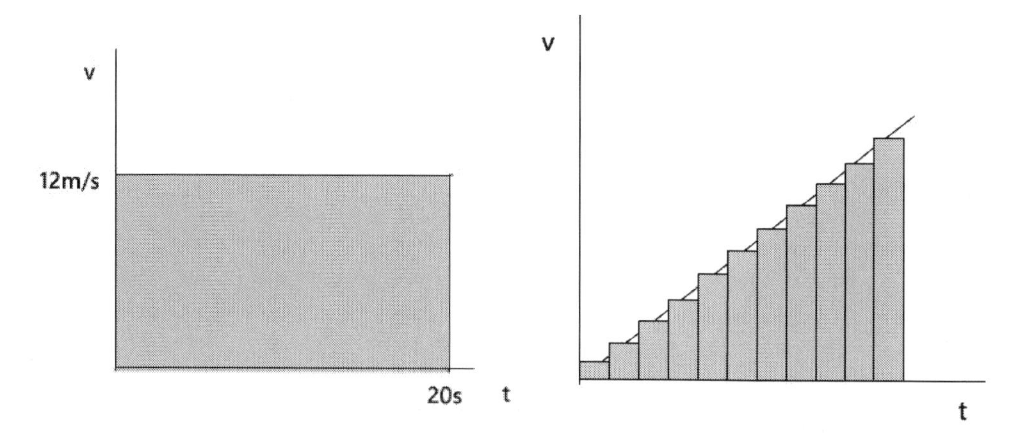

It should be clear from the first graph above that this is equal to the area of the rectangle.

Any graph can be split into rectangles with a base equal to a small fraction of time Δt

The distance travelled in each Δt would be the area of that rectangle so the total distance travelled would be the total area under the graph.

If the graph is a regular shape, such as a rectangle or a triangle, then finding the area is easy. E.g. for a triangle area = ½ base x height.

If it is an irregular shape then you may have to approximate the area or even count squares.

If we know the equation for v as a function of time then s can be found by integrating

i.e. $\mathrm{s} = \int \mathrm{v\,dt}$

Interesting that this equation is just $s = \sum v \times \Delta t$ (the sum of lots of triangle areas) for $\Delta t \rightarrow 0$

Acceleration from a velocity time graph

If the velocity of an object is changing then it is accelerating

$$\text{acceleration} = \frac{\text{change in velocity}}{\text{time}}$$

If an object is initially travelling with velocity u and then after time t it is travelling with a final velocity v then its average acceleration is given by

$$a = \frac{v - u}{t}$$

If the acceleration is uniform then this will equal the gradient of the graph.

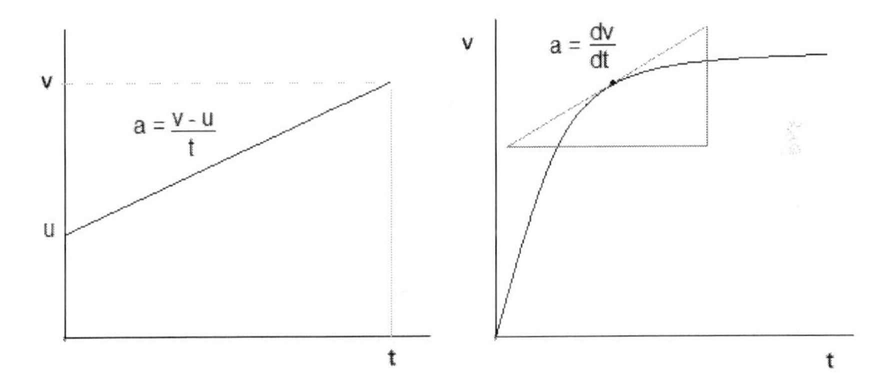

The acceleration is often not uniform, e.g. if you accelerate from rest in a car then after 30 seconds you will probably have reached a top speed.

If the acceleration is not uniform then we can find the instantaneous acceleration by finding the gradient of a tangent at that point.

The instantaneous acceleration is expressed as $a = \frac{dv}{dt}$

Remember

Displacement time graphs	Velocity time graphs
Gradient = velocity	Area =displacement Gradient = acceleration

Equations of motion

Quantity	Symbol	Units	Symbol for units
Displacement	s	Metres	m
Initial velocity	u	Metres per second	$m\ s^{-1}$
Final velocity	v	Metres per second	$m\ s^{-1}$
Acceleration	a	Metres per second squared	$m\ s^{-2}$
Time	t	Seconds	s

For situations where the acceleration is constant there are a set of very useful equations which we can use. These are called the equations of motion. Their nickname is the **s u v a t** equations which, if you look at the table above, should be obvious.

IMPORTANT: You can only use suvat when the acceleration is constant.

If it isn't then other methods must be used. E.g. if you were given a velocity time graph for an object reaching its terminal velocity, and you were asked to find the displacement. You would do it by estimating the area.

Do not worry too much about where these equations come from. The important thing for you is to learn them. Sometimes you may see them rearranged differently.

Equation	Derivation
$v = u + at$	This just comes from our definition of acceleration $a = \frac{v - u}{t}$
$s = ut + \frac{1}{2}at^2$	What would be the area under this velocity time graph? 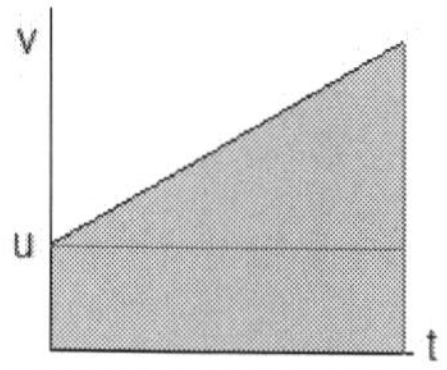
$\frac{s}{t} = \frac{v + u}{2}$	Two ways of working out the average velocity?
$v^2 - u^2 = 2as$	Substitute an expression for t from the first equation into the third

Using the equations of motion

1. *A car accelerates uniformly from rest at 3 m/s^2*

a) *What will be its velocity after 3.5s?*

b) *What distance will it have travelled?*

$v = u + at$ = 0 + (3 x 3.5) = 10.5 m/s

$s = ut + \frac{1}{2}at^2$ = 0 + (0.5 x 3 x 3.5^2) = 18.4m

2. *A child drops a coin into a well. 1.8s later they hear a splash. Take g = 9.8 ms^{-2}*

a) *How far did the coin fall?*

b) *What was its velocity when it hit the water?*

$s = ut + \frac{1}{2}at^2$ = 0 + (0.5 x 9.8 x 1.8^2) = 15.9m

$v^2 - u^2 = 2as$ as u = 0 $v = \sqrt{2\,g\,s}$ = 17.7 m/s

This assumes no air resistance, i.e. the acceleration is constant

3. *A train decelerates uniformly from 30m/s to 5 m/s over a distance of 1.5km*

a) *Calculate its deceleration*

b) *How long did it take?*

Average velocity = (30 + 5)/2 = 17.5 m/s so $t = \frac{s}{\text{average v}}$ = 85.7s

$a = \frac{v^2 - u^2}{2\,s} = \frac{5^2 - 30^2}{2 \times 1{,}500}$ = -0.583 m/s^2

Adding vectors

A vector is a quantity which has magnitude and direction. Scalar quantities only have magnitude.

Vectors	Scalars
Displacement	Distance
Velocity	Speed
Acceleration	Temperature
Force	Energy
Momentum	Mass

Vector quantities can be represented by drawing an arrow. The direction of the arrow represents the direction of the vector and, if drawn to scale, the size of the arrow represents the magnitude.

The sum of two or more vectors is called the **resultant**

If the vectors are in the same or opposite direction then the resultant is just the sum of the two vectors.

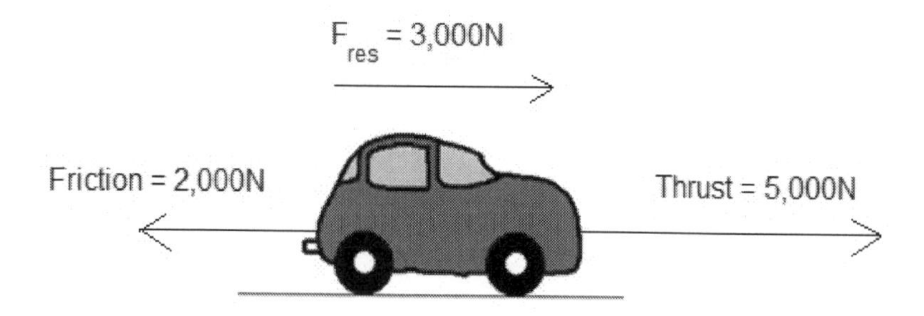

If the vectors are perpendicular then the magnitude of the resultant can be found using Pythagoras and the direction found using simple trigonometry.

2m/s

10 m/s

10m/s

2m/s

R

The diagram on the right is a vector diagram. To find the resultant we draw the vectors to be added, one after the other. The resultant is the single vector that would be the equivalent of these.

If it were drawn to scale you could actually measure the size of the resultant and the angle θ.

From Pythagoras $R^2 = 10^2 + 2^2$ so R = 10.2 m/s and $\theta = \tan^{-1}\frac{2}{10}$ = 11.3°

An aeroplane in still air flies at 100m/s. It aims north but a crosswind of 20m/s from the east blows it off course.

What will be its resultant velocity?

What direction should it aim in so as to travel north?

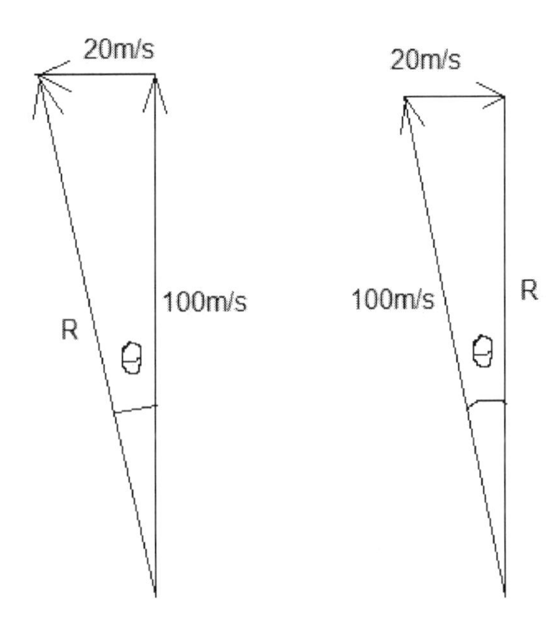

On the first diagram $R^2 = 100^2 + 20^2$ so R = 102 m/s $\theta = \tan^{-1}\frac{20}{100}$ = 11.3°

On the second diagram $R^2 = 100^2 - 20^2$ so R = 98 m/s $\theta = \tan^{-1}\frac{20}{98}$ = 11.5°

Be careful with the direction of the arrows

What if the vectors to be added are not in the same line or are not perpendicular?

We can still find the resultant from a scale drawing

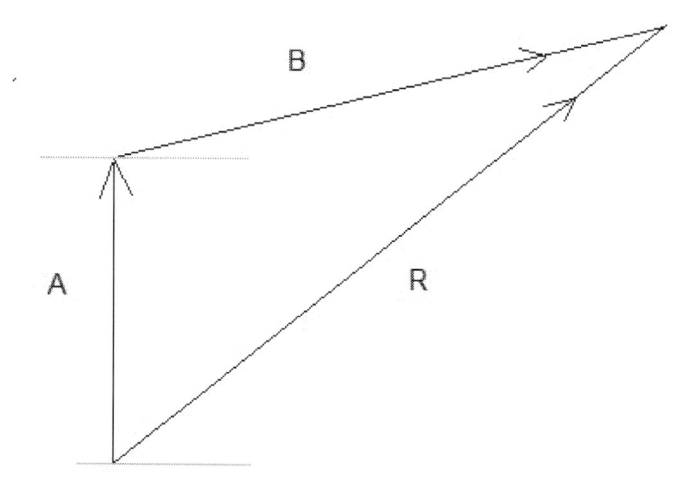

To get an accurate answer, however, you would need to be very careful with your drawing and it would need to be large.

There is a way of working out the resultant mathematically but we need to understand the next section first before we consider this.

Resolving vectors

In the diagram below the resultant force F is the sum of two perpendicular components F_x and F_y.

F_x is the component of F in the x direction F_y is the component of F in the y direction

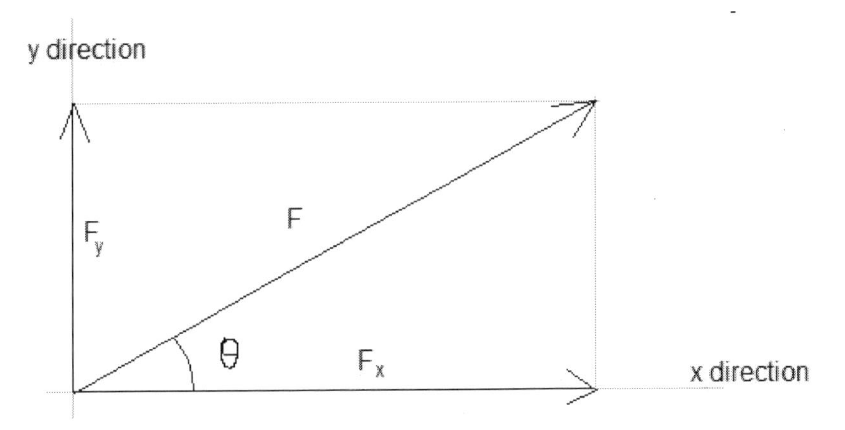

If we know the values of F and θ then we can calculate the components in the x and y directions using simple trigonometry

$$F_x = F\cos\theta \qquad F_y = F\sin\theta$$

The process of finding the components of a vector in two mutually perpendicular directions is called resolving and it is a very powerful and useful technique.

Consider these examples.

A hiker walks 3km in a direction 25° north of east

How far north and east of his starting point will he end up?

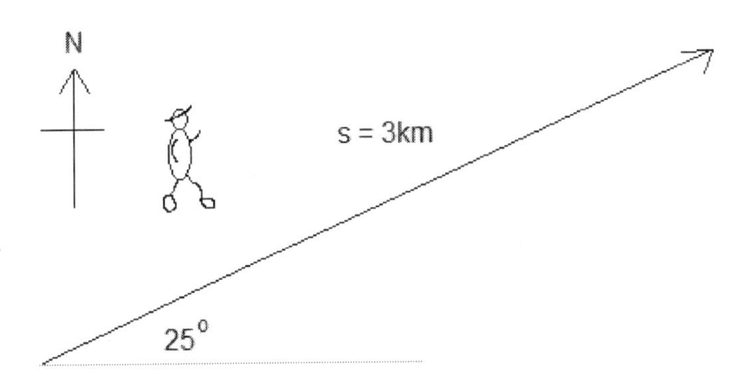

The component of s north will be $s\sin 25 = 1.27$ km

The component of s east will be $s \cos 25 = 2.72$ km

A toy car of mass 1.5kg rolls down a ramp which is at 30° to the horizontal

Calculate the components of its weight parallel to the slope and perpendicular to the slope.

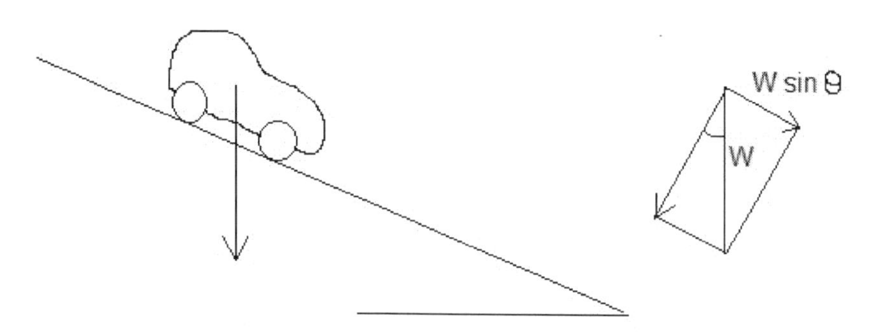

Notice that we can resolve in any two directions, as long as they are perpendicular. In this example we are resolving parallel to the slope.

Parallel $W \sin\theta = 1.5 \times 9.8 \times \sin 30 = 7.35N$

Perpendicular $W \cos\theta = 1.5 \times 9.8 \times \cos 30 = 12.7N$

Earlier I mentioned a mathematical way of adding two vectors together that were not perpendicular

Simply :

- resolve each vector into its x and y components
- find the sum of the x and y components
- you now have two perpendicular vectors which you can find the resultant of

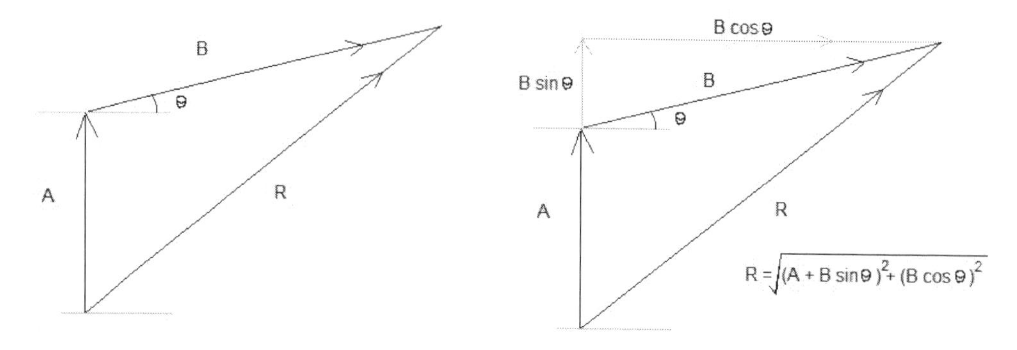

Newton's first and second law

In the sport of curling the player pushes the stone across the ice. As there is very little friction between the stone and the ice the stone does not slow down very quickly and travels a long way.

What if there was no friction? Would the stone slow down at all?

Newton's first law tells us the following:

1. If an object is stationary then there is no resultant force acting on it
2. If an object is moving at constant velocity then there is no resultant force acting on it

With no forces to slow it down a moving object would keep moving with the same velocity forever.

If all the forces acting on an object cancel and there is no resultant force then we say that the object is in **equilibrium**.

This can be a static equilibrium (when the object is stationary) or a dynamic equilibrium (when the object is moving with constant velocity).

Static equilibrium

Consider these examples of where an object is stationary

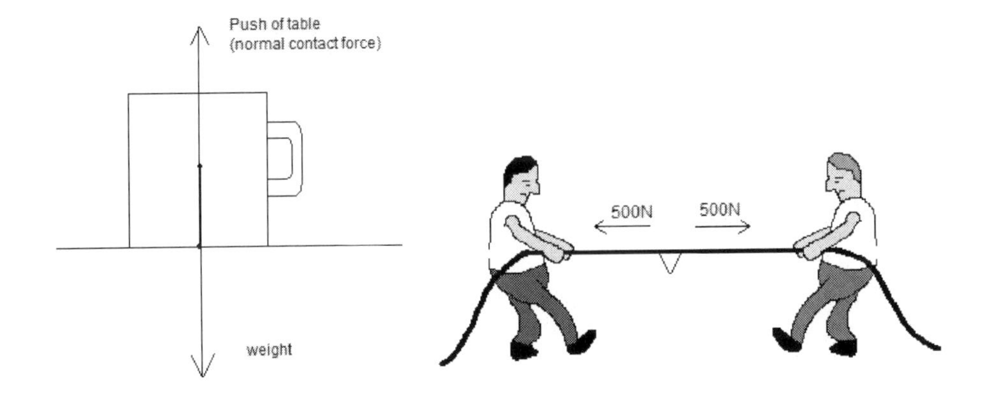

Dynamic equilibrium

Consider these examples where an object is moving with constant velocity

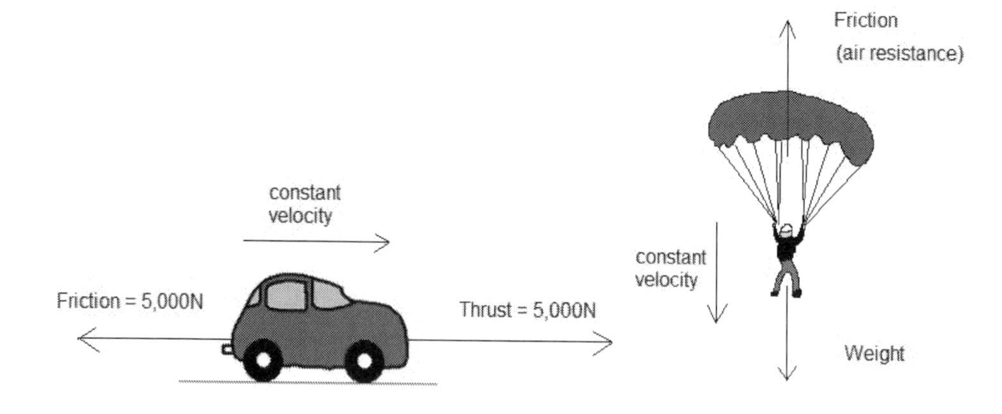

Newton's second law

What happens to an object when a resultant force does act on an object?

Newton's 2nd law says that it will produce a rate of change of momentum proportional to the force and in the same direction.

$$F_{res} \propto \frac{dmv}{dt}$$

This expression needs a bit of translating

- The momentum of an object is given by **momentum = mass x velocity**
- If a resultant force acts on an object then its momentum will change in the same direction
- How much the momentum changes every second will be proportional to the size of the force

In S.I. units (from the definition of the Newton) the equation becomes

$$F_{res} = \frac{dmv}{dt}$$

Now as $a = \frac{dv}{dt}$ we get the very useful expression $F_{res} = m\,a$

Note that I often include the subscript res to remind me that it is the **resultant** force that we use

A resultant force of 3,000N acts on a car of mass 900kg. Calculate its acceleration

$a = \frac{F}{m} \qquad = \frac{3{,}000}{900} \qquad = 3.33\ ms^{-2}$

The thrust force from a rockets motors = 500,000N at take off from Cape Canaveral. If the mass of the rocket is 20,000kg, calculate its acceleration at takeoff. *(g = 9.8 m/s²)*

$$a = \frac{F_{res}}{m} = \frac{500{,}000 - 196{,}000}{20{,}000} = 15.2\ ms^{-2}$$

Terminal velocity

When an object moves through a fluid, i.e. a liquid or gas, a resistive frictional force acts on it. This force increases with the velocity of the object.

This is why vehicles have a top speed and why sky divers reach a terminal velocity. As the resistive force increases then the resultant force decreases and, therefore, so does the acceleration.

The velocity time graph below is for a sky diver. Remember that the slope at any point tells us the acceleration.

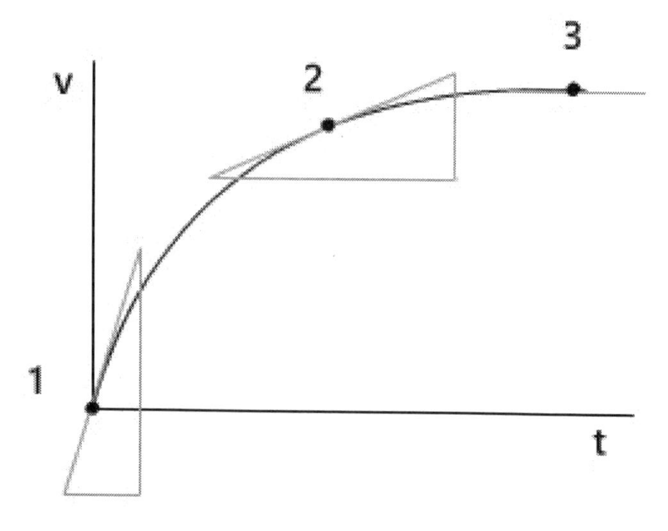

At point 1 the only vertical force is the sky divers weight	Acceleration = 9.8 m/s^2
At point 2 the resultant force = weight – drag	Acceleration = 5 m/s^2
At point 3 weight = drag so the resultant force = 0	Acceleration = 0

Work

James Joule was an English brewer. To begin with science was just a hobby and his initial discoveries about the nature of heat and mechanical energy came from his desire to build efficient equipment for his brewery.

Mechanical work is defined as follows:

Work is done by a force when the point of application of the force moves in the same direction.

Work = Force x distance $$W = F \times s$$

Work, like energy, is measured in Joules. In energy equations we will treat "work done" as being a form of energy although it is not something that an object would possess but the result of a process.

A man pulls a car with a force of 500N for 1.5 km. How much work does he do?

$W = F \times s = 500 \times 1{,}500 = 750kJ$

Note that the force and the displacement must be in the same direction for work to be done.

Notes:

- If there is no displacement then no work is done on the object. If you push a wall you are doing no work on the wall. The coffee cup on my desk pushes down on it but does no work.
- If the force is at an angle to the displacement then we need to resolve the force to find the parallel component, as in the example below.
- If the force is perpendicular to the displacement then no work is done. Gravity keeps the Moon in orbit around the Earth but does no work.
- "Work done by friction" is a useful concept which will appear in energy equations.
- Fields can do work. When an object falls gravity pulls it down so the gravitational field is doing work. When you lift an object you are doing work against gravity.

A horse pulls a barge as shown for 5km. How much useful work does it do?

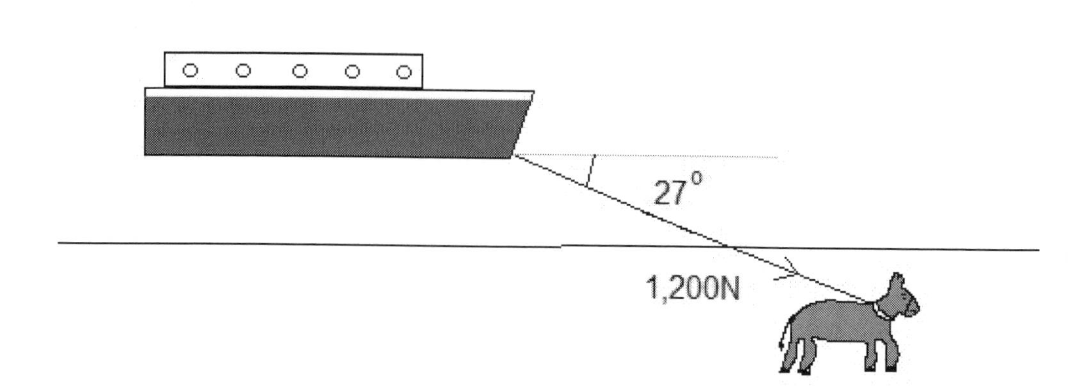

$W = F s \quad = 1.200 \cos 27 \times 5{,}000 \quad = 5.35 \text{ MJ}$

Kinetic Energy

Do not confuse kinetic energy with momentum. Kinetic energy is a scalar quantity and is measured in Joules. Momentum is a vector and is measured in kg m/s

The kinetic energy of a body is the energy a body possesses due to its movement. If work is done to accelerate an object then this will equal its gain in kinetic energy.

The equation can be derived from

$\mathrm{W} = \mathrm{F} \times \mathrm{s}$, $\mathrm{F} = \mathrm{m\,a}$ and $\mathrm{v}^2 - \mathrm{u}^2 = 2\mathrm{as}$

To calculate kinetic energy

$$K.E. = \frac{1}{2}\, m\, v^2$$

A dog of mass 20kg is running at 7m/s. Calculate its kinetic energy.

$\mathrm{K.E.} = \frac{1}{2}\,\mathrm{m\,v}^2 \quad = \frac{1}{2}\, 20 \times 7^2 \quad = 490\ \mathrm{J}$

A cannonball of mass 5kg has 36,000J of kinetic energy. Calculate its velocity.

$v = \sqrt{\frac{2 \times K.E.}{m}} \quad = 120\ \mathrm{m/s}$

Gravitational Potential Energy

Potential energy is the energy possessed by an object due to its position in a field of force.

An object which is higher up has more g.p.e. than one which is lower down. To say how much g.p.e. an object has at a certain height we would need to define the height at which it would have zero. To avoid this it is useful just to consider the change in g.p.e. in any given situation.

$\Delta G.P.E. = m\,g\,\Delta h$ where Δh is the difference in heights

This equation can be derived by considering how much work would need to be done to lift a mass m a height Δh.

A girl of mass 60kg climbs a hill 50m high. How much g.p.e. does she gain?

$\Delta G.P.E. = m\,g\,\Delta h$ $= 60 \times 9.8 \times 50$ $= 29.4$ kJ

Elastic Potential Energy

This is another type of energy that we should be familiar with which may crop up in mechanics questions.

If we have an object, such as a spring, with stiffness k then work must be done to deform it. If the object behaves elastically then this energy is stored in it as elastic potential energy and is recoverable.

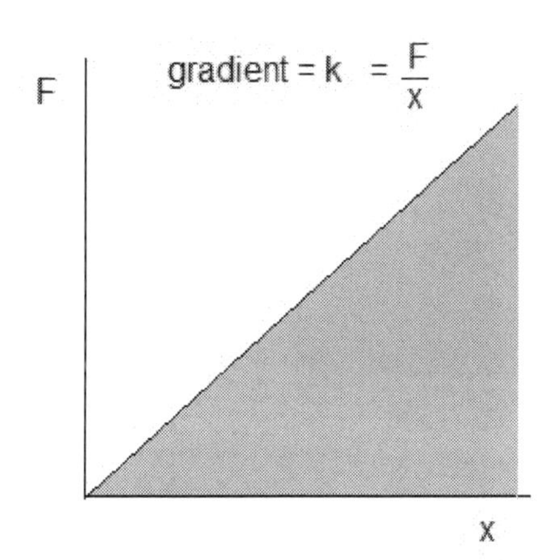

Area = energy stored $= \frac{1}{2} F x$

The force needed to stretch the object will increase as it is stretched so the total work done will be the average force multiplied by the extension

$$\text{E.P.E.} = \frac{1}{2} F x \quad = \frac{1}{2} k x^2$$

A spring extends 3cm when a force of 50N is applied to it

Calculate:

a) *Its stiffness*
b) *How much energy it stores*

$$k = \frac{F}{x} = \frac{50}{0.03} = 1{,}667 \text{ N/m}$$

$$\text{E.P.E.} = \frac{1}{2} F x \quad = 0.5 \times 50 \times 0.03 \quad = 0.75\text{J}$$

Conservation of energy

The principle of conservation of energy is one of the most important in physics. It will crop up in so many areas including thermodynamics, nuclear physics and electromagnetic machines. Here we are concerned with its application in mechanical problems.

Energy cannot be created or destroyed. It can only be transferred from one form to another.

Very often the best way to solve a problem is to use an "energy equation". If you identify the types of energy before and after an event and put them equal to each other. There are often other ways, e.g. suvat equations (when the acceleration is constant), but it's always nice to have options.

A man of mass 80kg dives off a cliff 20m high into the sea. Calculate his velocity when he hits the water.

<table>
<tr>
<td>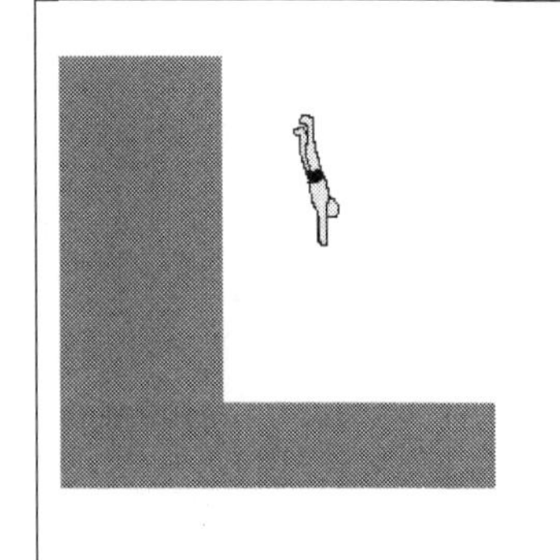</td>
<td>Our energy equation is
G.P.E → K.E.

So $\frac{1}{2} m v^2 = m g h$

re-arranging gives
$v = \sqrt{2 g h}$ = 19.8 m/s</td>
</tr>
</table>

An 80g pinball is launched horizontally with a velocity of 9m/s by a spring of stiffness 700N/m.

By how much was the spring compressed?

<table>
<tr>
<td>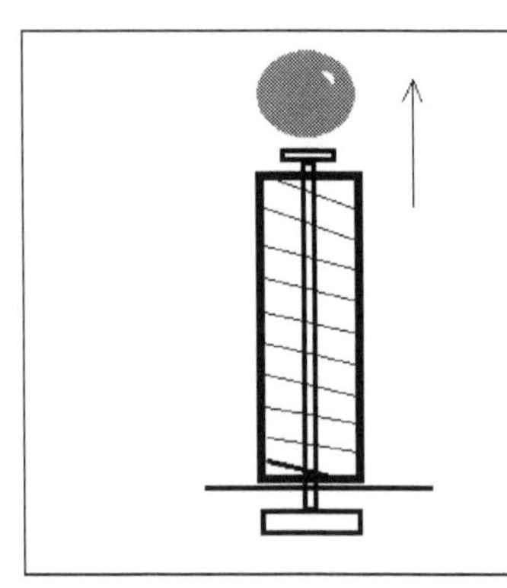</td>
<td>Our energy equation is
E.P.E. → K.E.

$\frac{1}{2} k x^2 = \frac{1}{2} m v^2$

So $x^2 = \frac{m v^2}{k}$ which gives x = 9.6 cm</td>
</tr>
</table>

A lady of mass 60kg skis down a hill 30m high which has a slope of 20°. She reaches the bottom travelling at 15m/s.

Calculate the average friction force acting on her as she skis down the slope.

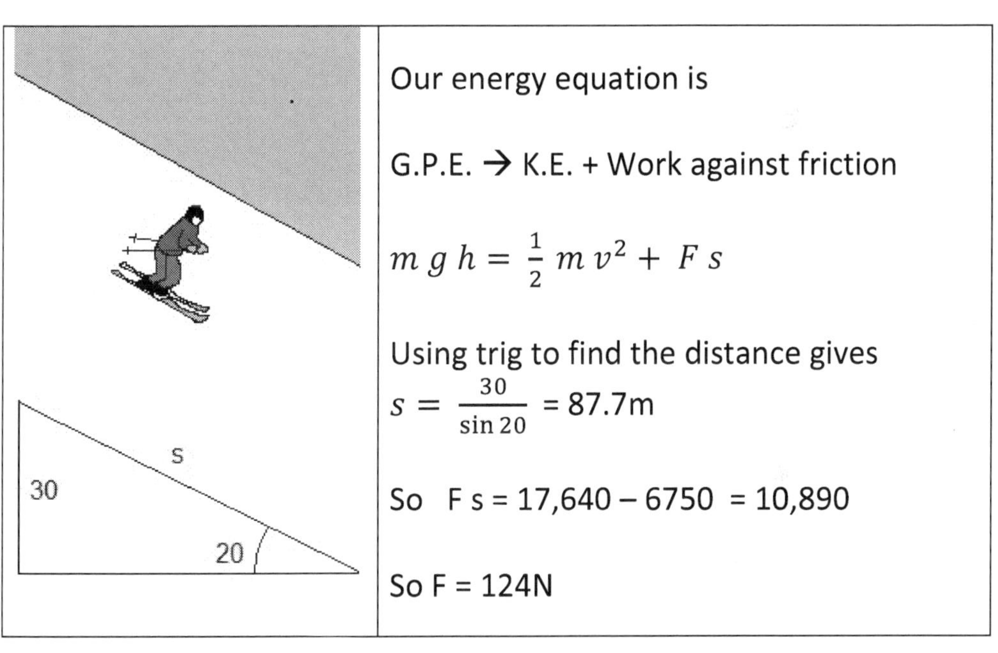

Our energy equation is

G.P.E. → K.E. + Work against friction

$$m\,g\,h = \frac{1}{2}\,m\,v^2 + F\,s$$

Using trig to find the distance gives

$$s = \frac{30}{\sin 20} = 87.7\text{m}$$

So $F\,s$ = 17,640 – 6750 = 10,890

So F = 124N

Notice how "Work against friction" can be a term in our energy equation

We are calculating the average force as the actual force at any instant will very much depend on the skier's velocity

Conservation of Momentum

Momentum was briefly mentioned before when we were discussing Newton's 2nd law.

Momentum = mv and is measured in kg m/s

Knowing the momentum of an object isn't actually very useful. What is useful is to consider the total momentum of a system of objects. (A system is a group of objects upon which no external forces act)

The total momentum of a system before an event must equal the total momentum after the event.

Momentum is conserved. The principle of conservation of momentum is another incredibly important and useful concept throughout physics.

An "event" in mechanics is usually an explosion or a collision.

In the collision above the total momentum of the two trolleys just before the collision **must** equal their total momentum just after the collision.

Remember that momentum is a vector so it can be positive or negative.

Example 1.

A trolley of mass 5kg travelling at 3m/s collides with a trolley of mass 2kg travelling in the opposite direction at 2m/s. The two trolleys stick together.

Calculate their velocity after the collision

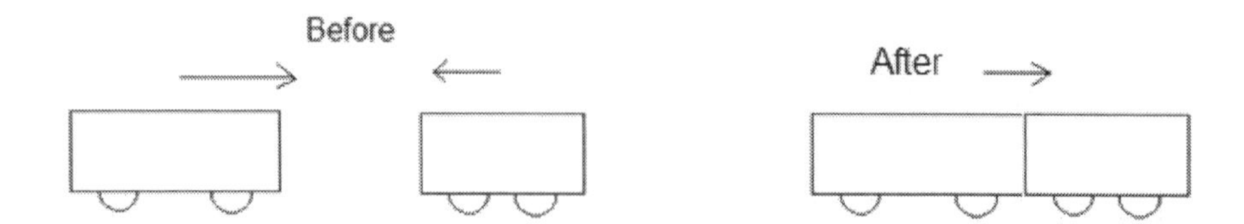

Momentum before = Momentum after

$(5 \times 3) - (2 \times 2) = (5 + 2) \times v$ so $v = 1.57$m/s

Example 2

A cannon of mass 300kg fires a cannonball of mass 2kg at a velocity of 200m/s

Calculate the recoil velocity of the cannon

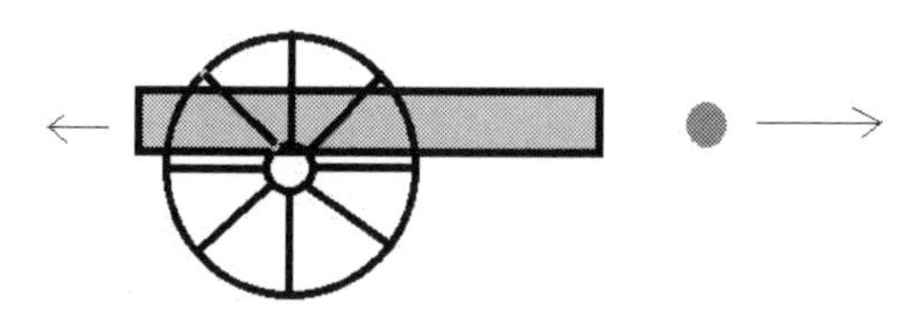

Momentum before = Momentum after

$$0 = (200 \times 2) - (300 \times v) \quad \text{so } v = -1.33\text{m/s}$$

Elastic and inelastic collisions

Momentum is always conserved in a closed system but is kinetic energy?

If you calculate the kinetic energy of the trolleys before and after the collision it is clear that kinetic energy is not conserved. Much of the initial kinetic energy is transferred into other forms, usually mostly internal energy. This is an inelastic collision.

If you drop a tennis ball it will bounce a few times but each time it collides with the floor it will lose some of its kinetic energy.

There are situations where the collision will be elastic. Usually these are ones where non-contact forces are involved, e.g. magnetic or electrostatic forces.

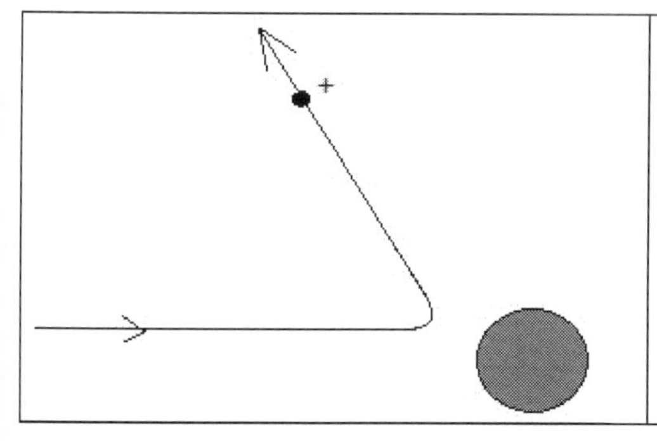	This diagram represents an elastic collision between an alpha particle and a gold nucleus. There is no loss of kinetic energy before and after the collision. When we model the behaviour of gases we assume that the collisions of gas particles are elastic.

Impulse

From Newton's 2nd law we have $F = \frac{\Delta mv}{t}$

Re-arranging this gives $F\,t = \Delta mv$

The quantity F t (which can be measured in Ns) is called the impulse of a force and so

Impulse = Change in momentum

When a driver crashes the function of the airbag is to increase the time it takes for the driver to slow down. By increasing the time for the collision the average force on the driver is less for the same change in momentum.

Below is a typical force against time graph for a collision

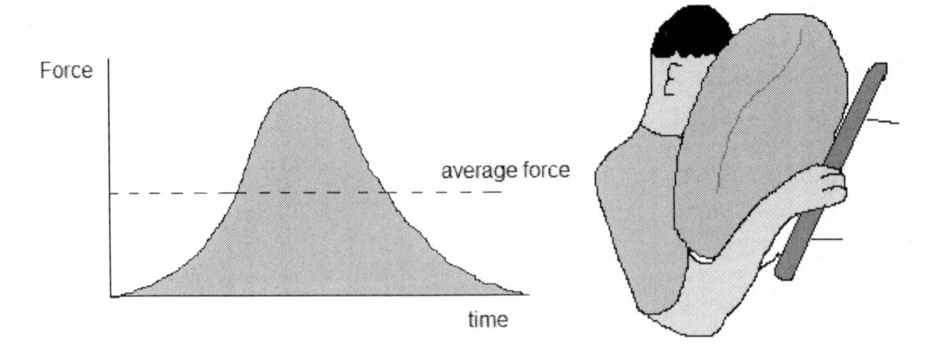

The area under the graph is equal to the change in momentum of the object. If the time for a collision can be increased then the graph would be flatter so the average, and maximum, forces would be less.

Momentum and propulsion

A rocket works by ejecting gases at a very high velocity. We can calculate the thrust produced using the equation $F = \frac{\Delta mv}{t}$

<table>
<tr>
<td>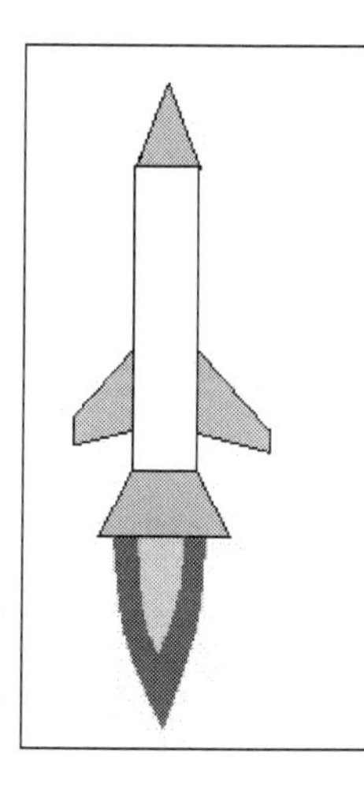</td>
<td>

A rocket ejects 4kg of exhaust every second at a speed of 400m/s

What thrust does this produce?

$$F = \frac{\Delta mv}{t} = \frac{4 \times 400}{1} = 1{,}600N$$

</td>
</tr>
</table>

An interesting point:

When two objects collide Newton's 3rd law tells us that the forces they exert on each other must be equal and opposite. As they must also act for the same time this means that the impulses of these forces must also be equal and opposite and so also their change in momentum.

If the momentum lost by one object must equal the momentum gained by the other object then it follows that **the total momentum of the system must be constant**, i.e. momentum is conserved.

Power

Power is rate at which work is done or the rate at which energy is transferred

$$P = \frac{W}{t}$$

Power can be measured in Joules per second (J/s) but is usually measured in Watts (W)

Both men below can do the same amount of work but the one on the left will do it in less time because he is more powerful.

A conveyer belt lifts 120kg of coal every minute a height of 10m into a furnace.

a) Calculate its useful power.
b) If its motor uses 600W of electrical power calculate the efficiency of the process

Ignoring any kinetic energy given to the coal

Work done by motor every second = gain in GPE of coal every second

$= (m\, g\, \Delta h) / 60 \quad = (120 \times 9.8 \times 10)/60 \quad = 196\text{W}$

$$\text{Efficiency} = \frac{\text{useful power output}}{\text{total power input}} = 0.326$$

A useful equation

Work = force x distance. The engine of a vehicle travelling at a steady speed does a certain amount of work every second. Well the distance it travels every second is equal to its velocity so

$$P = F\,v$$

The engine of a car travelling at a steady velocity of 25m/s produces a thrust force of 5.4kN. Calculate its useful power.

$P = F\,v \quad = 5{,}400 \times 25 \quad = 135\text{kW}$

Projectiles

So far we have only considered motion in just one dimension. The motion of a projectile is two dimensional, both horizontal and vertical.

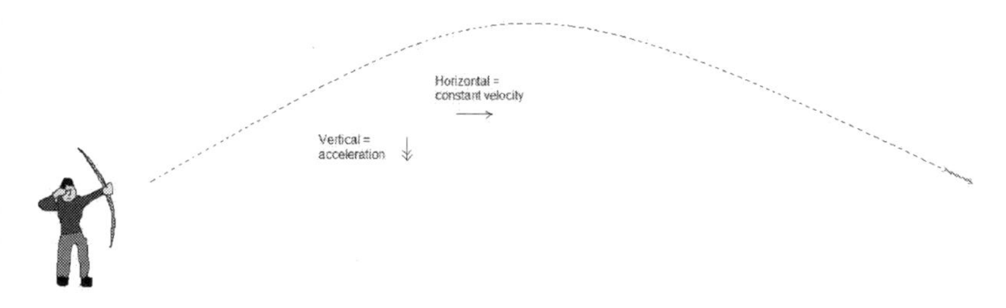

The trick, when dealing with projectiles, is to **deal with the horizontal and vertical components of their motion separately.**

Horizontally – ignoring air resistance the projectile moves with a constant velocity

Vertically – the projectile accelerates due to gravity, like any falling object

The combination of the two gives us a shape called a parabola (which actually means "path of a projectile"). It is interesting to consider how much scientific progress has resulted from research into warfare.

A car drives off a cliff horizontally at a velocity of 30m/s. The height of the cliff is 25m.

a) *How long will it take to hit the water?*
b) *What horizontal distance will it have travelled?*
c) *What will be its velocity when it hits the water?*

Vertically

$s = ut + \frac{1}{2}at^2$ as u = 0 re-arranging gives $t = \sqrt{\frac{2s}{g}}$ = 2.26s

Horizontally

$s = v\,t$ = 30 x 2.26 = 67.8m

<table>
<tr>
<td>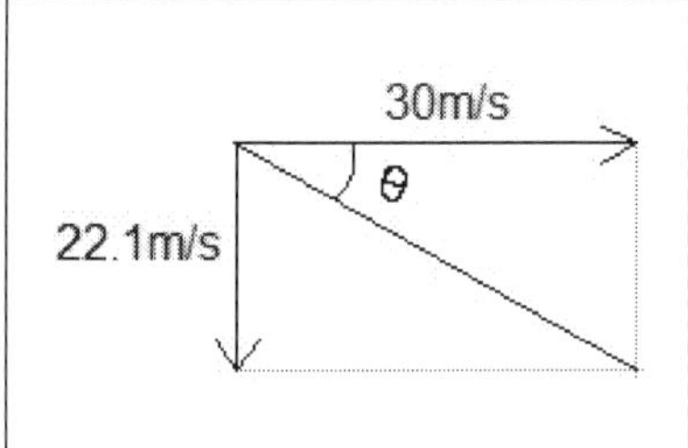</td>
<td>The vertical component of its velocity will be
$v = u + at$ = 9.8 x 2.26 = 22.1m/s

So its velocity will be
$\sqrt{30^2 + 22.1^2}$ = 37.3m/s
$\theta = \tan^{-1}\frac{22.1}{30}$ = 36.4°</td>
</tr>
</table>

Quantities

Quantity	Symbol	Units	Symbol for units
Displacement	s	Metres	m
Initial velocity	u	Metres per second	m s^{-1}
Final velocity	v	Metres per second	m s^{-1}
Acceleration	a	Metres per second squared	m s^{-2}
Time	t	Seconds	s
Momentum	p or mv	kilogram metres per second	kg ms^{-1}
Mass	m	kilograms	kg
Force	F	Newtons	N
Power	P	Watts	W
Work	W	Joules	J
Energy	E	Joules	J
Spring constant	k	Newtons per metre	N/m
Acceleration due to gravity	g	Metres per second squared	m/s^2

Equations to Learn

Average velocity $v = \frac{s}{t}$ Instantaneous velocity $v = \frac{ds}{dt}$

Average acceleration $a = \frac{v-u}{t}$ Instantaneous acceleration $a = \frac{dv}{dt}$

Suvat equations

$v = u + at$ $s = ut + \frac{1}{2}at^2$

$\frac{s}{t} = \frac{v+u}{2}$ $v^2 - u^2 = 2as$

Components of forces

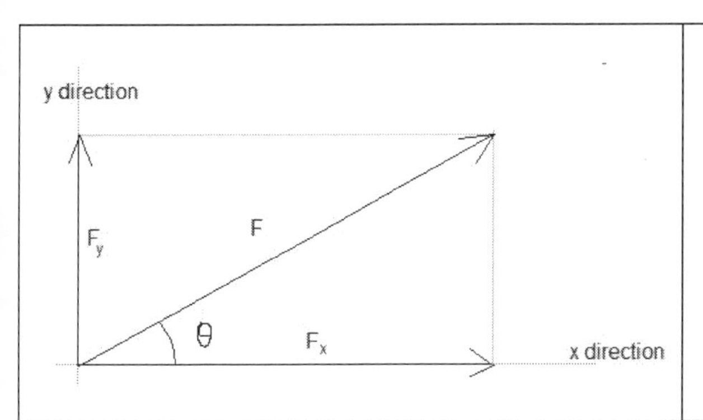

$$F_x = F\cos\theta$$

$$F_y = F\sin\theta$$

$$F^2 = {F_x}^2 + {F_y}^2$$

$$\tan\theta = \frac{F_y}{F_x}$$

Momentum = mv

$F = \frac{dmv}{dt}$ $F = \frac{\Delta mv}{t}$ $F = m\,a$

Impulse = F t = Δmv

$W = F\,s$

$P = \frac{E}{t}$ $P = \frac{W}{t}$ $P = F\,v$

Kinetic energy = $\frac{1}{2}mv^2$ Gravitational potential energy = $m\,g\,\Delta h$

Elastic potential energy = $\frac{1}{2}\,k\,x^2$

Questions

1. Here is a distance time graph for an object

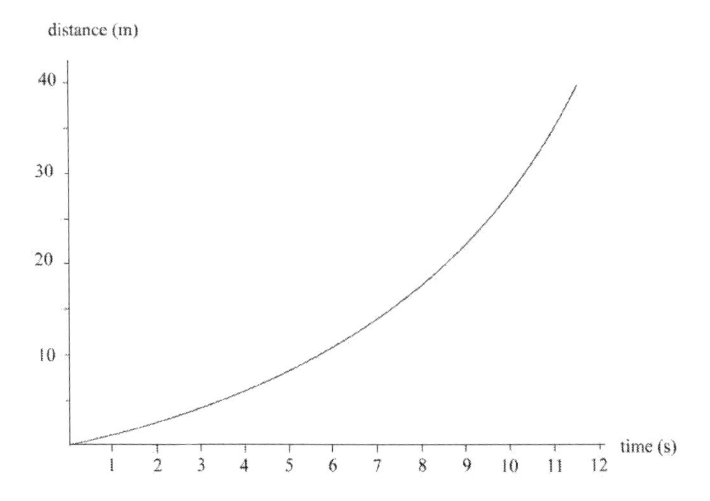

a) What distance has it travelled after 11 seconds?

b) What is its average speed in this time?

c) What is its speed at the start and at t = 11s?

d) Calculate its average acceleration in the first 11s

2. Here is a speed time graph for an object.

speed (m/s)

40
30
20
10

time (s)

1 2 3 4 5 6 7 8 9 10 11 12

a) What is the top speed of the object?

b) Estimate the distance travelled by the object in the time shown by the graph.

c) What is its average speed for the 12 seconds of the graph?

d) Find its acceleration a) at the start b) at $t = 4s$ c) at $t = 12s$

e) What object could this be a speed time graph for?

3. The friction force acting on a vehicle (mostly air resistance) is given by $F = Cv^2$ where C is a constant and v is its velocity

The engine of a certain car can generate a thrust force of 5,000N. The mass of the car is 1,200kg.

a) Calculate its initial acceleration assuming no friction is slowing it down

When the car is travelling at 10m/s the thrust force acting on it is equal to 900N

b) Use this information to calculate a value for C

c) Calculate the cars acceleration at this time

d) Calculate the top speed of the car

4. A man takes a running jump off a cliff. He is travelling at 7m/s horizontally when he leaves the cliff. Ignore effects of air resistance. Take $g = 9.8\ ms^{-2}$

a) What will his horizontal velocity be after 1 second?

b) Explain why

c) What will his vertical velocity be after 1 second?

d) Show the direction of his actual velocity after 1 second on a vector diagram and calculate its direction and magnitude.

5. An apple falls 3.4m from a tree to the ground. Take $g = 9.8\ m/s^2$

a) Calculate its speed when it hits the ground

b) Calculate the time it takes to fall

c) What assumption did you make in your calculation?

6. A train travelling at 45m/s decelerates uniformly in 80s and comes to rest

a) What distance does it travel in this time?

b) Calculate its deceleration

7. A toy boat which is powered by an electric motor travels at a speed of 2.4m/s in still water. It is put in a stream in which the water flows at 1.2m/s from left to right. It is aimed towards the opposite bank 4m away.

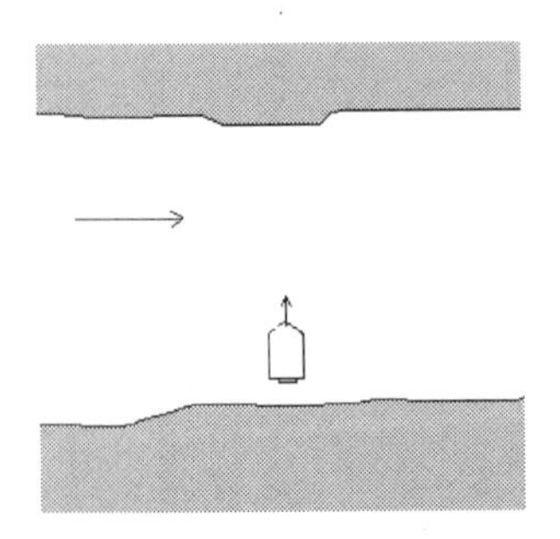

a) In what direction will the boat travel?

b) How long will it take to reach the opposite bank?

c) What will the actual speed of the boat be?

8. A shelf is attached to a wall by a hinge and kept horizontal by a wire attached to the wall. The directions of the forces acting on the shelf are shown below. The weight of the shelf is 50N.

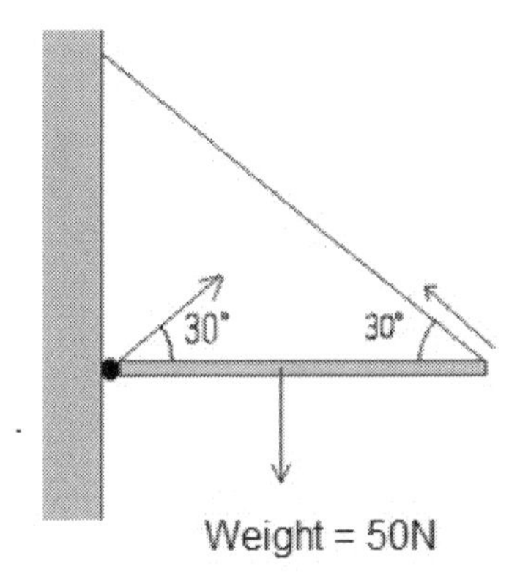

a) Identify the other 2 forces.

b) What resultant force acts on the shelf vertically?

c) Calculate how big the 2 unknown forces are.

9. A skier of mass 70 kg skis down a slope at an angle of 15^0 to the horizontal at a steady speed of 8 m/s.

a) What energy transfer is taking place?

b) What is her kinetic energy?

Calculate the following.

c) The horizontal and vertical distance she covers every second.

d) Her loss of g.p.e. every second.

e) The total friction force acting on her.

10. An air rifle pellet of mass 5g is fired into a lump of putty on a stationary trolley. The mass of the trolley and putty is 1.4kg. The trolley moves with a velocity of 0.6m/s after the impact.

a) Calculate the velocity of the pellet before the impact

b) Calculate the percentage loss of kinetic energy in the impact

Answers

1. a) 35m b) $35/11 = 3.18$m/s
   c) From gradients at start v = 1.6m/s at t = 11s v = 7.7m/s a = 0.56 m/s^2

2. a) 40m/s b) estimate area (e.g. big triangle) = 300m c) $300/12 = 25$m/s
   d. from gradients at start a = 20m/s^2 at t = 4s a = 3.8m/s^2 at t = 12s a = 0

A car accelerating from rest reaching a top speed

3. a) F =ma = 6,000 kN b) $C = F/v^2 = 9$ c) $a = (5{,}000 - 900)/1200 = 3.42$ m/s^2
d) if $v^2 = F/C$ = 5,000/9 v = 23.6 m/s

4. a) 7m/s b) $a_h = 0$ c) 9.8m/s

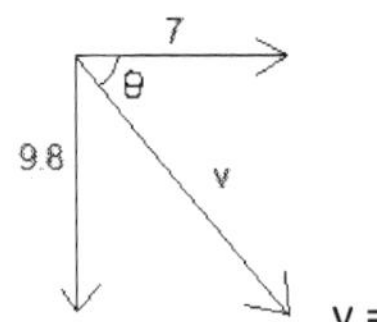

v = 12m/s θ = 54.4°

5. a) 8.16m/s b) 0.833s c) no air resistance

6. a) 1,800m b) 0.56m/s^2

7. a) 26.6° right of target b) 2.08s c) 2.68m/s

8. a) push of wall and tension b) 0 c) both 50N

9. a) GPE → Work against friction b) 2,240J c) H 7.73m/s V 2.07m/s d) 1,420J e) 178N

10. a) 169m/s b) 99.6%

Printed in Great Britain
by Amazon